LET'S LOOK AT LANDFORMS

THAT'S A VALLEY!

BY DWAYNE HICKS

Please visit our website, www.garethstevens.com. For a free color catalog of all our high-quality books, call toll free 1-800-542-2595 or fax 1-877-542-2596.

Library of Congress Cataloging-in-Publication Data
Names: Hicks, Dwayne, author.
Title: That's a valley! / Dwayne Hicks.
Other titles: That is a valley!
Description: New York : Gareth Stevens, [2022] | Series: Let's look at landforms! | Includes index.
Identifiers: LCCN 2020032282 (print) | LCCN 2020032283 (ebook) | ISBN 9781538263792 (library binding) | ISBN 9781538263778 (paperback) | ISBN 9781538263785 (set) | ISBN 9781538263808 (ebook)
Subjects: LCSH: Valleys–Juvenile literature.
Classification: LCC GB561 .H53 2022 (print) | LCC GB561 (ebook) | DDC 551.44/2–dc23
LC record available at https://lccn.loc.gov/2020032282
LC ebook record available at https://lccn.loc.gov/2020032283

Published in 2022 by
Gareth Stevens Publishing
29 E. 21st Street
New York, NY 10010

Designer: Tanya Dellaccio
Editor: Greg Roza

Photo credits: Cover Chandy Benjamin/Shutterstock.com; pp. 3–22 (background texture) MicroOne/Shutterstock.com; p. 5 Denis Belitsky/Shutterstock.com; p. 7 NiarKrad/Shutterstock.com; p. 9 Kath Watson/Shutterstock.com; p. 11 Guido Vermeulen-Perdaen/Shutterstock.com; p. 13 Nick Fox/Shutterstock.com; p. 15 Bob Hilscher/Shutterstock.com; p. 17 Nido Huebl/Shutterstock.com; p. 19 Joanna Rigby-Jones/Shutterstock.com; p. 21 Sina Ettmer Photography/Shutterstock.com.

Printed in the United States of America

CPSIA compliance information: Batch #CSGS22: For further information contact Gareth Stevens, New York, New York at 1-800-542-2595.

CONTENTS

Boldface words appear in the glossary.

What's a Valley?

A valley is a low area between hills or mountains. They are long but **narrow**. Most valleys have a river flowing through them. Many towns and cities are located in valleys. Valleys are shaped by **erosion** over thousands of years.

V-Shaped Valleys

Some valleys were made by rivers flowing over flat land. The river cuts a path into the ground and carries dirt and rock away. In time, this creates a V-shaped valley with **steep** sides. V-shaped valleys are called canyons.

Flat-Floored Valleys

V-shaped valleys grow wider as rivers continue to wear away the steep walls. This results in a wide, flat-floored valley. This is the most common kind of valley. Flat-floored valleys are common places for towns, cities, and farms.

Floodplains

The Nile River Valley in Africa is a flat-floored valley. Like other flat-floored rivers, the Nile has a floodplain. This is a wide, flat area of land near the river. The floodplain is covered with water when **floods** occur.

U-Shaped Valleys

Glaciers are huge masses of ice that form in cold mountains. They get bigger over time. A glacier "flows" very slowly. It scoops out land as it moves. When a glacier **melts**, it leaves behind a U-shaped valley.

Hanging Valleys

Sometimes a small glacier meets a larger glacier. When the weather warms and the glaciers melt, the smaller glacier creates a hanging valley. This is a small U-shaped valley high above a larger valley. They sometimes have waterfalls!

Rift Valleys

Earth's **surface** isn't one piece. It's made up of pieces called "plates." The plates move very slowly. When two plates move away from each other, a space called a rift appears between them. Over time, the rift grows into a valley.

The East African Rift System

The East African Rift **System** used to be called the Great Rift Valley. But it's actually a system of rift valleys. These valleys formed because two plates are moving away from each other. This also creates earthquakes and volcanoes!

Valleys and People

Valleys were good places for people to first settle down many years ago. Rivers gave people food and water. People built boats to make travel easier. Floodplains have rich soil for farming. Mountain valleys offer land for raising livestock.

GLOSSARY

erosion: the wearing away of the earth by wind or water

flood: a large amount of water covering an area of land that is usually dry

melt: to change from a solid (such as ice) to a liquid (such as water) when heated

narrow: long but not wide

steep: almost straight up and down

surface: the outer layer of something

system: a group of parts that are considered one whole thing when put together

FOR MORE INFORMATION

BOOKS

Amstutz, Lisa J. *Valleys.* North Mankato, MN: Capstone, 2020.

Hutchison, Patricia. *Valleys.* Mendota Heights, MN: North Star Editions, 2018.

WEBSITES

Great Rift Valley Facts for Kids

kids.kiddle.co/Great_Rift_Valley

Learn a lot more about the Great Rift Valley in East Africa.

Valley Facts for Kids

kids.kiddle.co/Valley

Read more about valleys and other landforms at this comprehensive website.

INDEX